BUILDING GREENER, RESILIENT TRANSPORT INFRASTRUCTURE

INNOVATING WITH ARTIFICIAL INTELLIGENCE AND DIGITAL TWINS IN ROAD DESIGN IN UZBEKISTAN

SEPTEMBER 2023

ASIAN DEVELOPMENT BANK

Contents

Tables, Figures, and Boxes

Tables

Figures

Boxes

Foreword

Building greener and more climate-resilient infrastructure is crucial today due to the catastrophic impacts of climate change. In Central Asia these impacts include water scarcity, desertification, and increased frequency of extreme weather events such as droughts and floods. Since transport infrastructure is particularly vulnerable to these effects, resilient road networks that can withstand climate change impacts are necessary to ensure the continued functioning of essential services and protection even during adverse climatic conditions.

Artificial intelligence (AI) and other digital solutions have immense potential in building greener and more climate-resilient infrastructure. They can assist in detecting and predicting extreme weather events, optimizing resource usage, and decarbonizing infrastructure projects along their life cycle. By leveraging the power of AI and digital tools, more efficient and sustainable infrastructure can be created, with mitigation and adaptation plans in place.

I am therefore pleased to introduce this publication, produced in collaboration with ORIS, which provides a case study on the upgrade of Uzbekistan's A380 highway (section km 673–698).

The multidimensional insights generated from using AI and digital twins for the project will be of interest to policymakers, practitioners, and researchers working at the intersection of climate change and transport infrastructure.

I am confident that this publication will be of great benefit to a wide range of stakeholders. I hope that it inspires further research and action to mitigate the impacts of climate change on transport infrastructure in Central Asia.

Chen Chen
Director
Central and West Asia, East Asia, and the Pacific Team
Transport Sector Office
Sectors Group
Asian Development Bank

Acknowledgments

This publication shares learning from a collaborative road design assessment. Asian Development Bank (ADB) staff from multiple departments worked with construction materials platform ORIS and government partners to optimize the design of a section of highway in Uzbekistan.

The project was supported by ADB's Digital Innovation Sandbox and was piloted by the Transport and Communications Division of the Central and West Asia Department.

The ADB team included Pawan Karki, a principal transport specialist in the Sectors Group, supported by former director Hideaki Iwasaki; Marc Lepage, a principal IT specialist (Technology Innovation), Information Technology Department; and internal reviewers Michael Anyala (senior road asset management specialist) and David Shelton (senior transport specialist, road safety).

ORIS conducted the digital due diligence, and its CEO Nicolas Miravalls led the preparation of this publication. ORIS team members included Hugo Pley-Leclercq, Koji Negishi, Ann Somera, and Elodie Woillez.

ADB also appreciates the contributions of external reviewer Adélaïde Feraille at the Ecole des Ponts ParisTech.

This project was undertaken with generous support from the People's Republic of China Poverty Reduction and Regional Cooperation Fund and the Republic of Korea e-Asia and Knowledge Partnership Fund.

Abbreviations

ADB	Asian Development Bank
CAREC	Central Asia Regional Economic Cooperation
CO_2eq	carbon dioxide equivalent
FSI	fatality and serious injury
IPCC	Intergovernmental Panel on Climate Change
iRAP	International Road Assessment Programme
kg	kilogram
km	kilometer
kt	kiloton
LCA	life cycle assessment
mm	millimeter
UN	United Nations

Executive Summary

The aim of this publication is to share knowledge on the opportunity of conducting early due diligence in infrastructure projects to support the Sustainable Development Goals. It provides a case study on the potential offered by innovative digital tools to facilitate advanced impact assessment. It aims to share learning that can support similar projects in different contexts and geographies.

The case study looks at the design of a project in Uzbekistan to upgrade part of the A380 highway (section 673–698 km). In assessing the pavement design, a team from the Asian Development Bank partnered with ORIS to pilot a multiple-scenario digital analysis. The project demonstrates how road design can be significantly improved by making informed and data-driven decisions at an early stage, based on local conditions. The following box summarizes the key characteristics of the A380 highway upgrade design.

Uzbekistan's A380 Highway Upgrade Design

Key characteristics of the upgrade of section 673–698 km of the A380 highway in Uzbekistan

A **25-kilometer (km) highway section** upgrade to widen it from one lane to two lanes in each direction, with a separated carriageway;

Objective: **reduce the travel time by 15%** while **accommodating 25% more daily traffic**; and

Ambition to design the road upgrade in a sustainable manner, with an **optimized carbon footprint**; **a lower use of natural resources, including water**; a **cost-efficient design**; **resilience to climate change**; and a **reduction in fatalities by 25%** along the section.

INITIAL DESIGN: The base case design was a rigid pavement structured with six layers of materials, including one cement concrete pavement, three layers of crushed stones, and two layers of nonwoven geotextiles.

The assessment of the base case design showed the following impact measurements:

Volume of materials used:
909,000 tons

Carbon footprint:
39,800 tons carbon dioxide equivalent

Water consumption:
86.2 million liters

Pavement costs:
$30.06 million

Source: ORIS. 2022. Digital Twin Road Solution Draft Report: Uzbekistan (A380 Section 1). Unpublished.

Through the ORIS Materials Intelligence platform, the project was digitized to create a digital twin. After compiling and processing local construction materials into the ORIS platform, a comparative analysis on multiple scenarios was conducted, using digital simulations and artificial intelligence. This advanced digital technology allowed a quick assessment around a few key indicators (as shown in the ensuing table): (i) pavement solutions and mechanical performance over the life cycle; (ii) circular economy options; (iii) carbon emissions and environmental impact; (iv) resource consumption, including water; (v) safety prediction based on international assessment methods of the International Road Assessment Programme; and (vi) resilience to climate change via evaluation of the main project risks (such as heat, freeze–thaw cycle, water runoff floods, landslides, and silting).

Assessment of Key Road Indicators

Road Pavement Design	Road Safety Analysis	Road Sustainability	Road Resilience
Improvement opportunities identified from pavement design analysis: • Pavement and material cost reduction of $3.2 million, • Reduction of 11.9 kilotons (kt) of natural resources, and • Circular economy with 165,000 tons of recycled materials.	For a noticeable improvement on predicted safety performance of the road, invest $7.1 million to exceed the United Nations safety target of 3 stars and prevent over 1,800 fatalities or serious injuries.	Save 6.79 kt of carbon dioxide equivalent (a 17% saving from the base design) and 4.8 million liters of water via mitigation measures. Smart highway innovations can improve safety and resilience of the infrastructure for an initial investment of $2.2 million.	To help the road withstand heat, freeze-thaw cycles, and water runoff flooding, invest $19.4 million in climate change adaptation and mitigation measures. This would help reduce damage and avoid early heavy repairs.

Source: ORIS. 2022. Digital Twin Road Solution Draft Report: Uzbekistan (A380 Section 1). Unpublished.

Using artificial intelligence allows quick data collection and interactive alternative design analysis. During this 11-week project in 2022, the team gathered data on geography, construction materials, pavement solutions, and climate to develop a first set of conclusions. These conclusions enabled the project managers to better understand the best design options adapted to the local context and in line with their sustainability goals.

Comparison with the initial A380 highway upgrade design using the platform showed that alternative designs had a better sustainability impact.

The project shows the potential of digital twin platforms to develop a comprehensive and innovative approach to road pavement design. This approach can support quick due diligence and can facilitate investments in lower-carbon, safer, circular, and more resilient road networks, in line with the Sustainable Development Goals.

I The A380 Highway Project: An Opportunity for a Multidimensional Design Assessment

Road networks remain valuable assets, supporting efficient transportation of people and goods. Roads help people access trade, jobs, education, and health-care facilities, supporting growth and improved living standards.

To achieve the sustainability goals, road construction designs must evolve to integrate the climate crisis, resource scarcity, and social inequalities, using new digital technologies. Each road has a unique local environment because of its traffic and weather conditions, along with the availability of local sources of materials. The design choice is a major factor on its costs (60% of total construction costs) and its carbon footprint (85% of embedded carbon emissions). Road investments can be better monitored, thanks to new tools using digital calculations associated with data management and artificial intelligence.

A. Upgrade of the A380 Highway in Uzbekistan

The Asian Development Bank (ADB) partnered with ORIS, an intelligent materials platform, to perform due diligence on the design of a 25-kilometer (km) section of A380 highway in Uzbekistan (section 673–698 km).

The A380 highway in Uzbekistan is 1,204 km long and is part of the Central Asia Regional Economic Cooperation (CAREC) corridor. ADB has been supporting the rehabilitation of the CAREC corridor in Uzbekistan through multiple projects, in collaboration with the Uzbekistan Committee of Roads. The A380 highway section 673–698 km is a section spanning 25 km from Turtkul to Nukus (Figure 1), to be upgraded and widened from one lane to two lanes in each direction, with a separated carriageway. These works are directed by the Government of Uzbekistan with the support of ADB.

The project aims to support better connectivity and regional cooperation, while improving the road condition and safety, and designing a sustainable and resilient road. It aims to decrease fatalities along the section by 25% and reduce travel time.

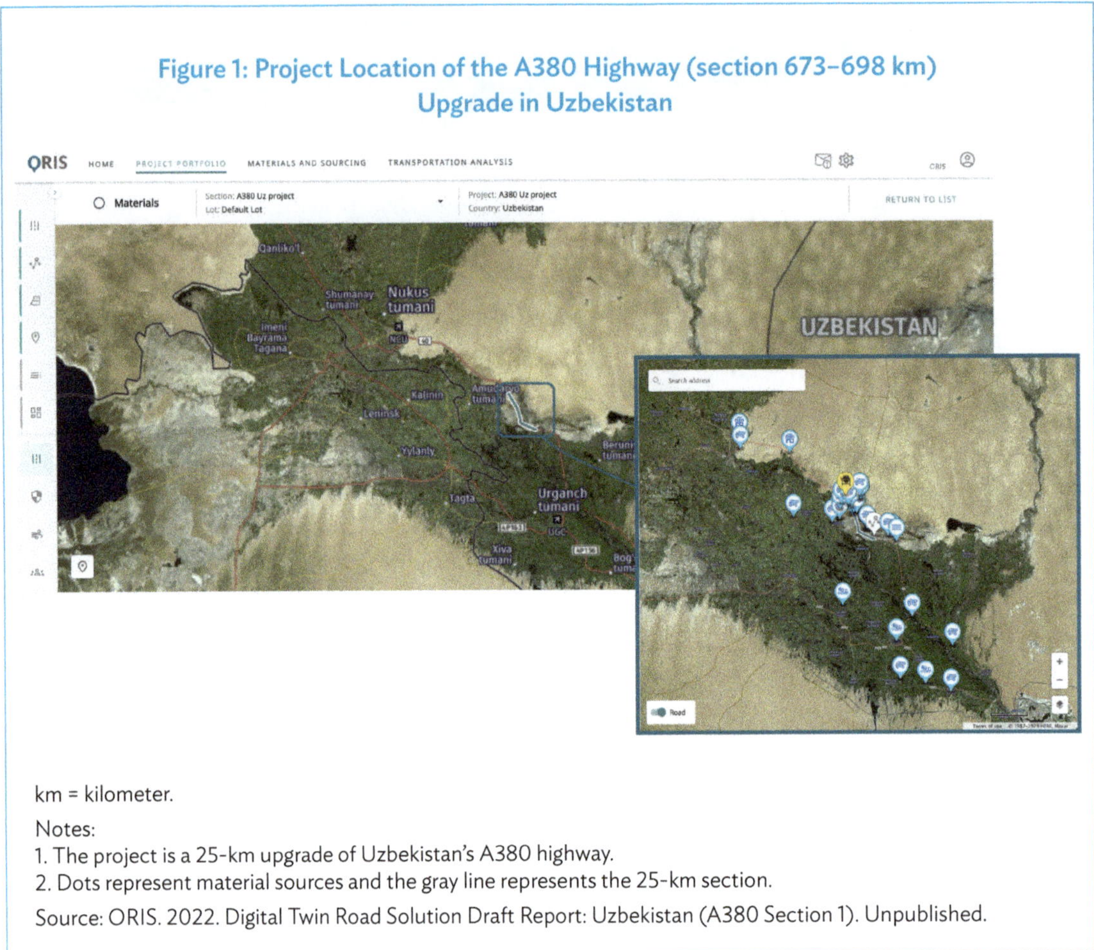

km = kilometer.
Notes:
1. The project is a 25-km upgrade of Uzbekistan's A380 highway.
2. Dots represent material sources and the gray line represents the 25-km section.
Source: ORIS. 2022. Digital Twin Road Solution Draft Report: Uzbekistan (A380 Section 1). Unpublished.

B. Project Objectives

Working with ORIS Materials Intelligence, the ADB team was able to set up a digital twin. Digital twins are virtual models designed to accurately reflect a physical object.[1] The object being studied replicates vital areas of functionality to produce data about different aspects of the physical object's performance. Once informed with such data, the virtual model can be used to run simulations, study performance issues, and generate possible improvements—all with the goal of generating valuable insights. A digital twin is a virtual environment, which makes it considerably richer for study than simulations. The difference between a digital twin and a simulation is largely a matter of scale. While a simulation typically studies one particular process, a digital twin can itself run any number of useful simulations to study multiple processes.

To build this digital twin into the ORIS platform, the project data was integrated along with its real environment characteristics: local sourcing, traffic, and weather conditions. Local pavement norms were also part of the data input.

---

[1]   IBM. What Is a Digital Twin? https://www.ibm.com/topics/what-is-a-digital-twin.

Using this digital twin, the ADB team was able to assess multiple pavement design options along the following lines:

- Analysis of initial pavement design on costs, carbon emissions, resilience, and resource consumption;
- Maintenance scenario for durability and resilience to climate change to find the optimum design for the use phase and anticipate extreme climate conditions;
- A safety due diligence on designs and alignments (vertical and horizontal) to provide recommendations on road safety, considering geometry, alignment, visibility, traffic, and various other road safety parameters; and
- Incorporation of hydrological calculations for drainage structures to design optimum sections of structure and waterway openings (road hydraulic transparency), in case of peak water flows and floods.

Through these multiple simulations, the team was able to identify the optimum design options that best fit its criteria.

Box 1 briefly describes ORIS Materials Intelligence.

Box 1: ORIS—A Digital Materials Platform Based on Artificial Intelligence

ORIS Materials Intelligence is the first construction materials platform for the smart use of resources and low-impact infrastructure, based on materials data knowledge and sharing. Using construction materials data knowledge, ORIS Materials Intelligence is able to assess the impact of linear infrastructure designs in a multidimensional view. The platform allows to compute parameters and data, such as materials properties, geolocation, expected traffic, and weather conditions, for an effective and sustainable infrastructure construction.

ORIS Materials Intelligence is a deep-technology innovation at the crossroads of several areas of expertise. It combines advanced technologies such as data science, infrastructure engineering, and artificial intelligence or machine learning, which all require a high level of technical expertise and specialized knowledge to use effectively.

The data science component of ORIS Materials Intelligence involves collecting, cleaning, and analyzing large datasets about the environmental impact of different materials. The infrastructure engineering component involves understanding the technical properties of different materials and how they behave under different conditions. This requires a deep understanding of material science, structural engineering, and other specialized fields, as well as the ability to use advanced modeling and simulation tools to predict the performance of different materials in different scenarios. The artificial intelligence or machine-learning component of ORIS Materials Intelligence involves using advanced algorithms and machine-learning techniques to make predictions and recommendations based on the data and information collected by the platform. This requires expertise in programming, machine learning, and artificial intelligence, as well as the ability to use specialized tools and frameworks to develop and implement these algorithms.

Source: ORIS Materials Intelligence. www.oris-connect.com.

Methodology
Project's Digital Modeling Using Artificial Intelligence Based on Local Data

A. Data Collection

The first phase of the project was to incorporate into the ORIS database local materials sourcing capabilities and standards.

ORIS Materials Intelligence combines data from multiple sources, transforming and enriching them to offer an exhaustive and qualitative cartography of production sites (Figure 2). The combination of geospatial data and machine learning is used to detect and classify the production sites, ensuring a fast acquisition of data and an optimal quality. The final manual verification of data guarantees an optimal accuracy and supports continuous improvement of the artificial intelligence models.

The sourcing database was built using satellite recognition, artificial intelligence, and ORIS Materials Intelligence detection algorithms to identify the 75 sites available for the project across Uzbekistan, with a focus around the A380 highway section 673–698 km: 44 quarries and borrow pits, nine concrete plants, seven cement factories, six asphalt plants, six train and/or road multimodal areas, two water sources, and one geotextile factory.

The site location, type of site (cement factory, quarry, etc.), and type of materials produced were the key attributes detected with the geospatial data and machine learning. The site survey then confirmed this information and added the materials pricing, equipment, and energy used to produce the materials (factors for the calculation of the carbon footprint).

The combined use of satellite recognition, artificial intelligence, and ORIS algorithms enabled the project team to (i) allocate more time on the project refinement and studies, and less time on material resources research at the early stage; and (ii) benefit from a more granular analysis of the project challenges at an early stage.

The pricing of materials was incorporated based on a local survey conducted in November 2021, local experiences, and references for similar projects.

In a local context where suppliers have not produced an Environmental Product Declaration for their materials—meaning that the carbon emissions are unknown—ORIS Materials Intelligence compiles local and international data to define the carbon footprint for each material. Based on those initial carbon evaluations, ORIS calculates the overall impact of the construction using its recognized methodology.[2]

---

[2] ORIS Materials Intelligence. 2023. ORIS Materials Intelligence Receives Intertek Assurance Statement on Its Carbon Footprint Calculation Models for Infrastructure Projects. Media release. https://oris-connect.com/en/news/media-release-oris-receives-intertek-assurance-statement-on-its-carbon-footprint-calculation-models-for-infrastructure-projects.

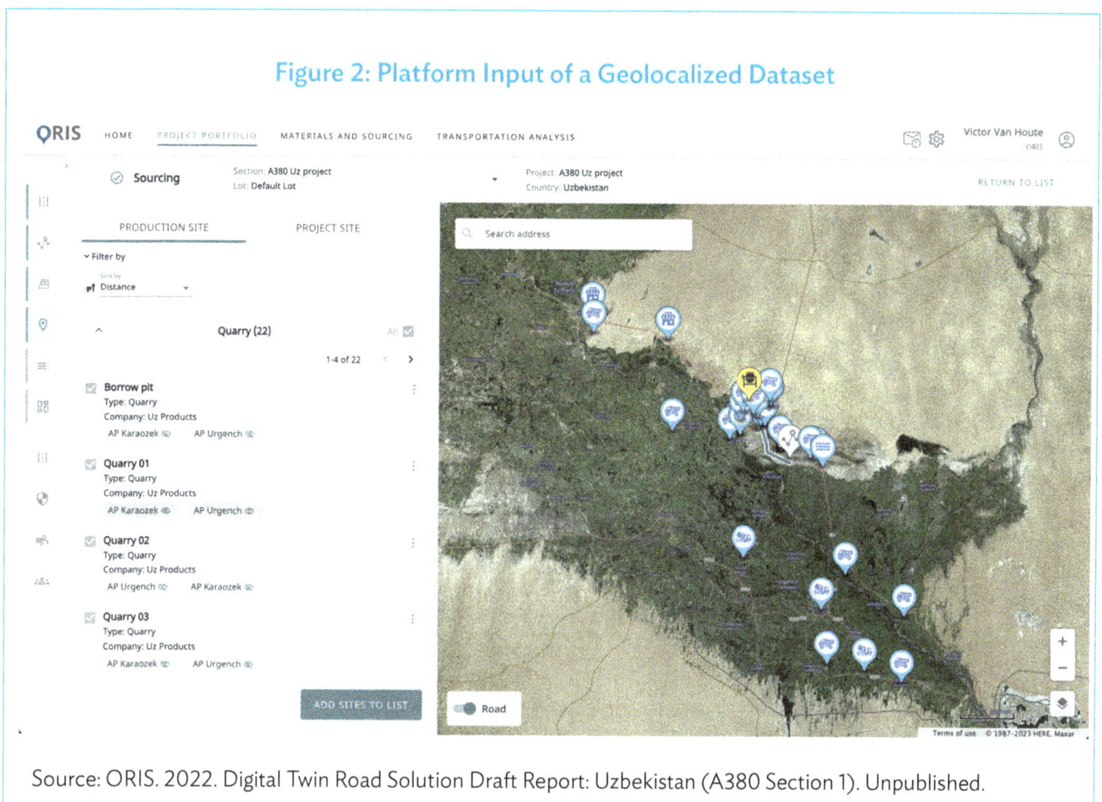

Source: ORIS. 2022. Digital Twin Road Solution Draft Report: Uzbekistan (A380 Section 1). Unpublished.

For instance, the cement carbon assessment was based on local data from two local cement plants and completed with the tool calculation of the Global Cement and Concrete Association, with a default hypothesis taken for the region of the Commonwealth of Independent States.[3] The calculation was based on a life cycle assessment (LCA) model—from the extraction of primary materials to the factory gate—that considered Uzbekistan's local energy mix for manufacturing. The Gabi database provided background datasets for the modeling of other materials such as asphalt-based materials, fuels, aggregates, product transportation to a project jobsite, and waste treatment.[4] The ORIS Materials Intelligence platform compiled those data to provide the carbon emission factor of the cement produced in each of the two plants. This was then used in the ORIS Materials Intelligence database with the other material carbon factors to calculate the overall carbon impact of the 25 km section.

The locally available materials and design regulations were gathered from the following Uzbekistan standards regulation, in addition to international road design standards:
- Material standards: GOST[5]
- Pavement design: ADB Handbook I for cement concrete road, Tashkent, 2021; SETRA 98, French pavement design catalog 1998; and RStO 12, German guidelines for the standardization of pavement structures of traffic areas, 2012

---

[3] Global Cement and Concrete Association. Environmental Product Declarations. https://gccassociation.org/sustainability-innovation/environmental-product-declarations/.

[4] Sphera. Product Sustainability Software and Data. https://gabi.sphera.com/international/databases/.

[5] FSBI Institute for Standardization. https://www.gostinfo.ru/.

B.   Project Conditions for Modeling

The next phase of the project was to input data specific to the project and its conditions to develop the most appropriate model.

The base case design provided by the Uzbekistan Road Committee was a rigid pavement structured with six layers of materials:
  (i)   270 millimeters (mm) cement concrete pavement B30, with sulfate-resisting Portland cement;
  (ii)   Nonwoven geotextile, 450 grams per square meter;
  (iii)   180 mm crushed stone gravel sand mixture, reinforced with sulfate-resisting Portland cement;
  (iv)   150 mm crushed stone mixture C5;
  (v)   200 mm crushed stone mixture C5; and
  (vi)   Nonwoven geotextile, 200 grams per square meter.

To consistently compare with the base case, two alternative rigid structures were selected according to European standards (Figure 3):
  •   Alternative 1: This design is based on the French SETRA 98 pavement catalog, with a suppression of one layer of geotextile and the optimization of the concrete layer from 270 mm to 250 mm. To meet the expected performance, the aggregate layer thickness and cement treated base course layer thickness are increased.
  •   Alternative 2: This design is based on the RStO 12 German guidelines, with one layer of geotextile and the optimization of the base cement treated layer. To meet the expected performance, the aggregate layer thickness is increased.

Comparison has been made between rigid pavements only at the request of the Uzbekistan Road Committee to follow the presidential decree stipulating that all new highway road pavements should be in cement concrete.

To better anticipate how materials will perform under local weather conditions and over their lifetime, multiple material properties and factors were included in the model, particularly flexural strength of concrete, elasticity, fatigue behavior of concrete, and subgrade reaction.

Traffic conditions, based on studies conducted between 2017 and 2021, were also considered. The following assumptions were made:
  •   Design life: 30 years
  •   Average daily truck traffic per direction: 1,190
  •   Growth factor constant during the studied period: 5.5%
  •   Reference axle load type: 13 tons

As the project is located in a very dry area, water savings were identified as a key priority for the project. The use of water is crucial to reach the optimum moisture content in road pavements; however, during geotechnical inspections, groundwater was absent in the pits dug up to 3 meters in depth. Two water sources away from the site were subsequently considered for the study. Water consumption for pavement construction was calculated with the following assumptions:
  •   Optimum moisture content for aggregate layers and treated layers (percentage of the total mass), and
  •   Water for concrete layers in the mixed design (percentage of the total mass).

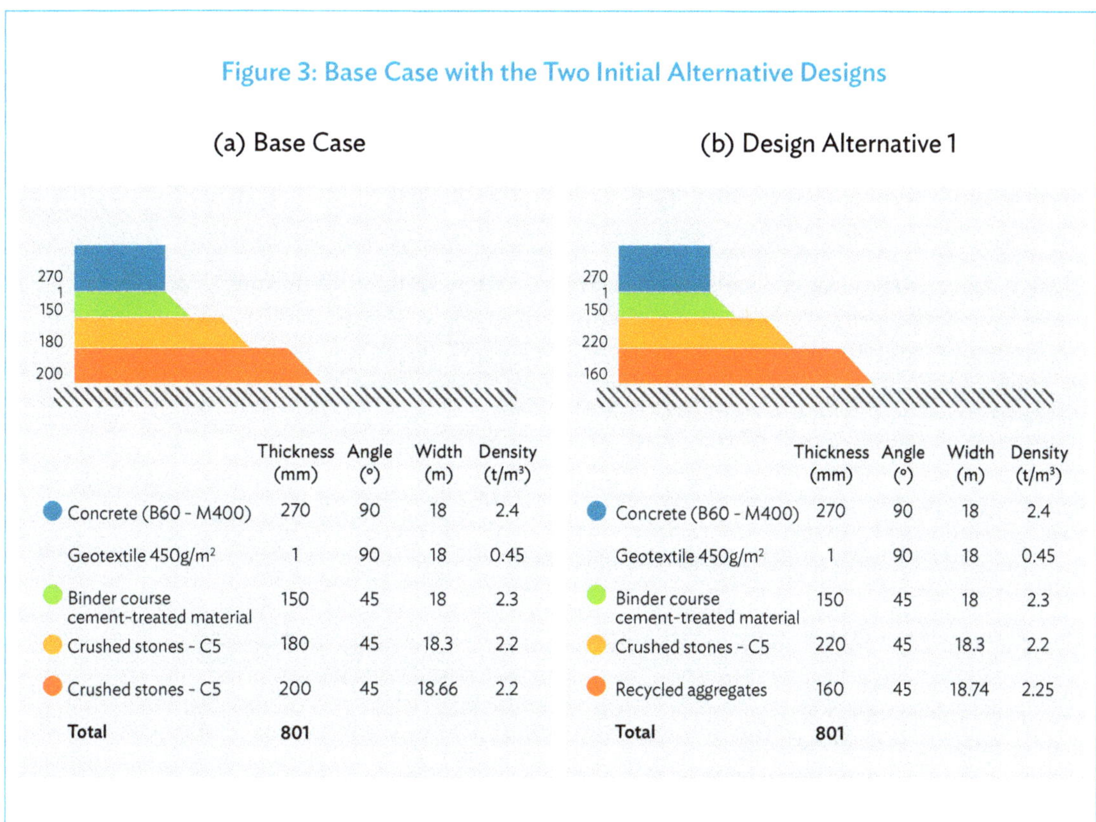

Figure 3: Base Case with the Two Initial Alternative Designs

(a) Base Case

	Thickness (mm)	Angle (°)	Width (m)	Density (t/m³)
Concrete (B60 – M400)	270	90	18	2.4
Geotextile 450g/m²	1	90	18	0.45
Binder course cement-treated material	150	45	18	2.3
Crushed stones – C5	180	45	18.3	2.2
Crushed stones – C5	200	45	18.66	2.2
Total	**801**			

(b) Design Alternative 1

	Thickness (mm)	Angle (°)	Width (m)	Density (t/m³)
Concrete (B60 – M400)	270	90	18	2.4
Geotextile 450g/m²	1	90	18	0.45
Binder course cement-treated material	150	45	18	2.3
Crushed stones – C5	220	45	18.3	2.2
Recycled aggregates	160	45	18.74	2.25
Total	**801**			

(c) Design Alternative 2

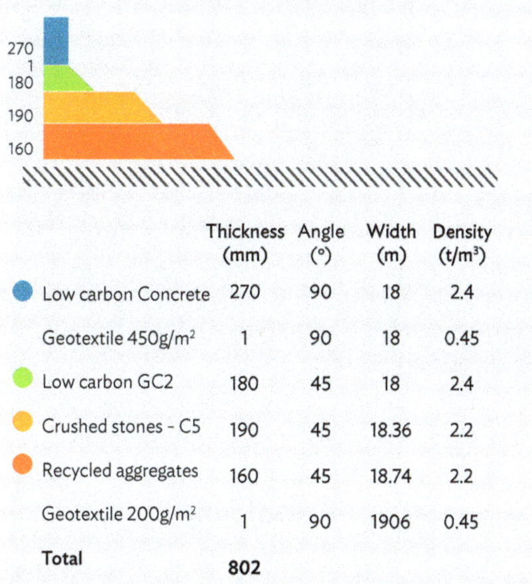

	Thickness (mm)	Angle (°)	Width (m)	Density (t/m³)
Low carbon Concrete	270	90	18	2.4
Geotextile 450g/m²	1	90	18	0.45
Low carbon GC2	180	45	18	2.4
Crushed stones – C5	190	45	18.36	2.2
Recycled aggregates	160	45	18.74	2.2
Geotextile 200g/m²	1	90	1906	0.45
Total	**802**			

° = degree, C5 = crushed stone type C5, g = gram, GC2 = cement-treated base, m = meter, m² = square meter, m³ = cubic meter, mm = millimeter, t= ton.

Source: ORIS. 2022. Digital Twin Road Solution Draft Report: Uzbekistan (A380 Section 1). Unpublished.

Through the modeling set up in the ORIS platform, the ADB team performed a pavement design analysis comparing the base case with the alternative designs. This analysis identified improvement opportunities to lower costs by $3.2 million, spare 14 kilotons (kt) of natural resources, and enable the use of 165 kt recycled materials.

A. Cost Calculation and Natural Resources Analysis Show Room for Improvement

The total cost of the pavement was shared among the different stages of construction:
- Product stage: raw material supply, transport, and manufacturing costs;
- Transport cost: includes the finished product transportation from production site to the project main access point;
- Internal transport cost: transportation of finished product from the main access point at the site to other work locations, as needed; and
- Construction installation process cost: application costs.

For a more accurate view, the modeling into the platform allowed the inclusion of a light maintenance scenario to prevent cracks and improve the International Roughness Index after 15 years of services as follows:
- In years 5, 10, and 15: crack sealing and join renewing (according to needs);
- In year 15: diamond grinding the concrete surface layer to bring back the optimal International Roughness Index; and
- In years 20 and 25: crack sealing and join renewing (according to needs).

Maintenance cost is estimated at $1.35 million over the 30-year service of the A380 highway section 673–698 km, representing 2.25% of the total project cost. The maintenance program is designed to have a limited budget and carbon impact.

The cost comparison of the base case with the selected alternatives identified that both present lower costs (–1.7% for the base case and –4.2% for the alternatives). Design alternative 2 presents a benefit of $2.2 million cost reduction compared to the base design with a similar performance.

Consumption levels of natural resources for the base case and the alternative designs were calculated according to the project section length, layer width, thickness, and densities. As the project is located in a water-scarce area, water was considered in the model to reflect the stress on material consumption. With a material consumption reduction of 2.0% compared to the initial base design, design alternative 2 would preserve 15 kt of materials compared to the base design with similar performances.

B. The Benefits of Using Recycled Materials from the Existing A380 Highway

Through this assessment, the use of recycled material from the existing A380 highway was identified as an opportunity for further optimization of the pavement design.

Design alternative 2 was analyzed with the use of a recycled layer of aggregates from the existing road. The existing section was composed of good quality materials (asphalt, crushed stones, shoulder materials, and a borrow pit), estimated at a volume of 165 kt.

Modeling of design alternative 2 with a recycled pavement shows it offers great performance in terms of costs and material consumption compared to the base case. The final savings amount to $3.2 million, representing 6.7% of the total project cost. Material consumption is reduced by 14 kt of virgin materials, while 165 kt of materials are reused. Figure 4 presents the material consumption depending on the design chosen.

The use of this recycled layer also has a positive impact on the following:
- Carbon performance. Indeed, the existing road (used as a subgrade) is a resource for materials with great performance and could be used as a granular foundation without major transportation or transformation. In addition, borrowed pit materials with lower carbon impact will be used to replace those recycled materials as subgrade materials instead of producing crushed rock materials; and
- Water consumption of the project (section 4).

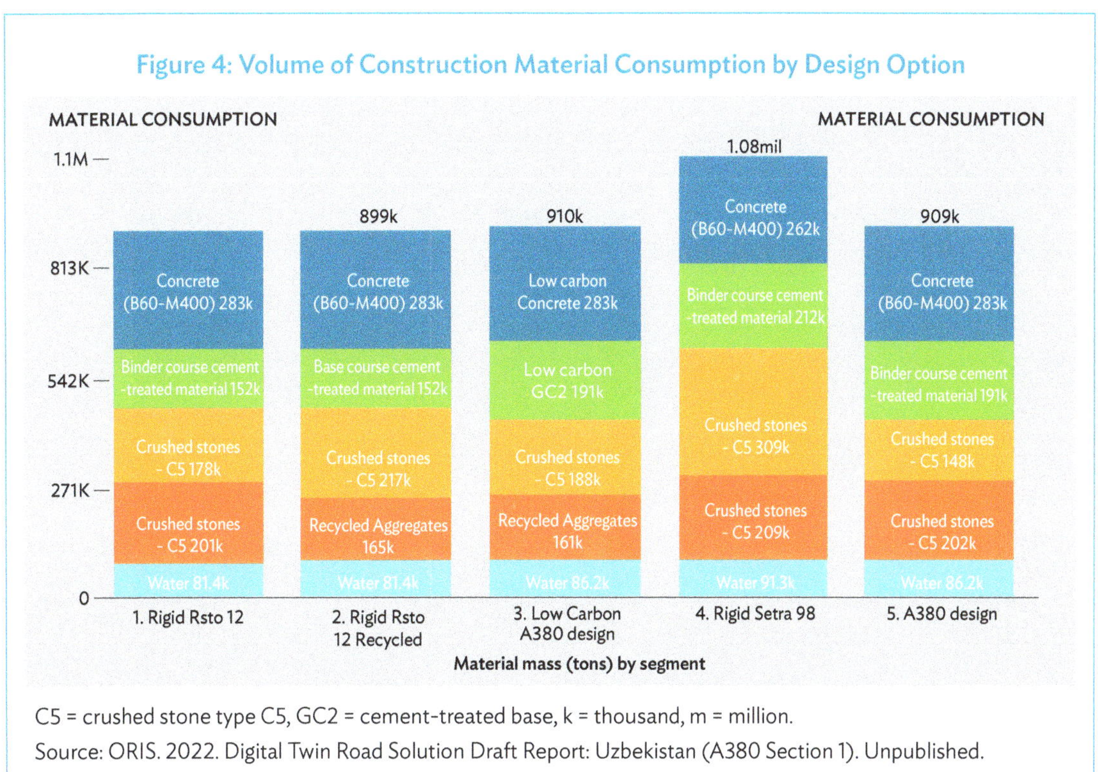

C5 = crushed stone type C5, GC2 = cement-treated base, k = thousand, m = million.

Source: ORIS. 2022. Digital Twin Road Solution Draft Report: Uzbekistan (A380 Section 1). Unpublished.

Aligning the Project's Carbon Performance with Climate Goals
Circular and Low-Carbon Solutions

By challenging the identified alternatives with low-carbon and circular solutions, the project's environmental footprint was significantly improved, with 6.69 kt of carbon dioxide equivalent (CO_2eq) avoided (16.8% compared to the base case) and 4.8 million liters of water saved.

A. Life Cycle Assessment of the Project

Life cycle assessment (LCA) is a methodology for assessing the environmental impacts of a product, service, or technology, considering all the stages of its life cycle, from raw material extraction all the way to demolition and end-of-life. It is a standard method widely used to assess the environmental impacts of products or systems. Using an LCA approach, which considers all factors impacting the carbon footprint over the lifetime, is recognized as enabling the most accurate measure of the carbon impact. Traditionally, a carbon footprint analysis in the infrastructure sector evaluates, in most cases, embodied carbon emissions, including material production and road construction phases—and, in the best case, the maintenance phase. However, the operational phase of the infrastructure would have a contribution as much as embodied carbon. This operational phase is mostly linked to energy and water consumption by system users. Besides, the end-of-life scenarios would give wide ramifications in decision-making because of the valorization of demolished waste. A carbon footprint analysis through a whole life cycle of a road helps identify carbon hotspots and make the right decisions.

The ORIS Materials Intelligence LCA is limited to the climate change indicator, material consumption, and water consumption.[6] In compliance with LCA standards,[7] the ORIS platform integrates the calculation of carbon emissions (CO_2eq) of a road pavement from the extraction of raw material to the end-of-life stages, including the following:
  (i)   product stage;
  (ii)   construction stage;
  (iii) use phase stage, which considers the carbonation, albedo effect processes, deflection, roughness, and texture impact on vehicle consumption;
  (iv)   maintenance phase;
  (v)   operational energy use; and
  (vi)   disposal of deconstruction-related waste.

Using the ORIS platform, the ADB team evaluated that carbon emissions from the base case cumulate to 39.2 million kilograms (kg) of CO_2eq for the product and construction phases, while the use phase's emissions cumulate to 15.7 million kgCO_2eq.

---

[6]   Basically, the global LCA is a multicriteria analysis including climate change, photochemical ozone creation, and acidification; whereas, ORIS Materials Intelligence primarily focuses on the climate change indicator, material consumption, and water consumption.

[7]   These standards include ISO 14040, ISO 14044, and EN 15084.

The comparison with design alternative 1 shows an increase of 2.2% after 30 years of service life compared to the base design, with similar values for design alternative 2. This implies that deflection, roughness, and texture impact on vehicle consumption for design alternative 1 is higher than the base design because of the pavement design.

The ORIS platform also quantified the emissions due to fuel consumptions of daily road transportation (cars and trucks) during the service lifetime of the road. It appears that the construction, maintenance, and use phases of the road only represent 3.7% of the total carbon emissions (CO_2eq), including the impact due to traffic over the 30-year service life of the infrastructure.

B. Reducing Carbon Emissions

As mentioned in section 3, the use of a recycled layer on the pavement of design alternative 2 reduced the carbon emissions by 2.6 million $kgCO_2$eq. To explore further reductions, the use of the platform enabled the identification of solutions based on international standards.

By investing in a local cement plant to reduce its carbon footprint, the materials supplied to the project would have a lower carbon footprint. The replacement of the concrete and treated bases with low-carbon products from hydraulic-blended material reduces the carbon emissions by 6.69 million $kgCO_2$eq.

The limited investment of $2.5 million would also support the decarbonization of the local industry. The final savings between this alternative design with recycled pavement and low-carbon products and the base case amounts to $2.3 million. This translates into a savings of 5.0% compared to the total project cost. Figure 5 summarizes the carbon emissions of the design options.

C. Reducing Water Consumption in a Water-Scarce Area

The A380 highway section 673–698 km project is located in a water-scarce area of the Republic of Karakalpakstan. The nearest source of water is located about 25 km from the construction site. Water monitoring was a priority for the project.

With a deep analysis of the local material properties and the material mix design, alternative pavements can bring a reduction of up to 4.8 million liters of water. The analysis was made based on the mix design for concrete and the need of water for the optimal compaction of the materials.

Appendix 1 provides a comparative summary between the base case and the design alternatives.

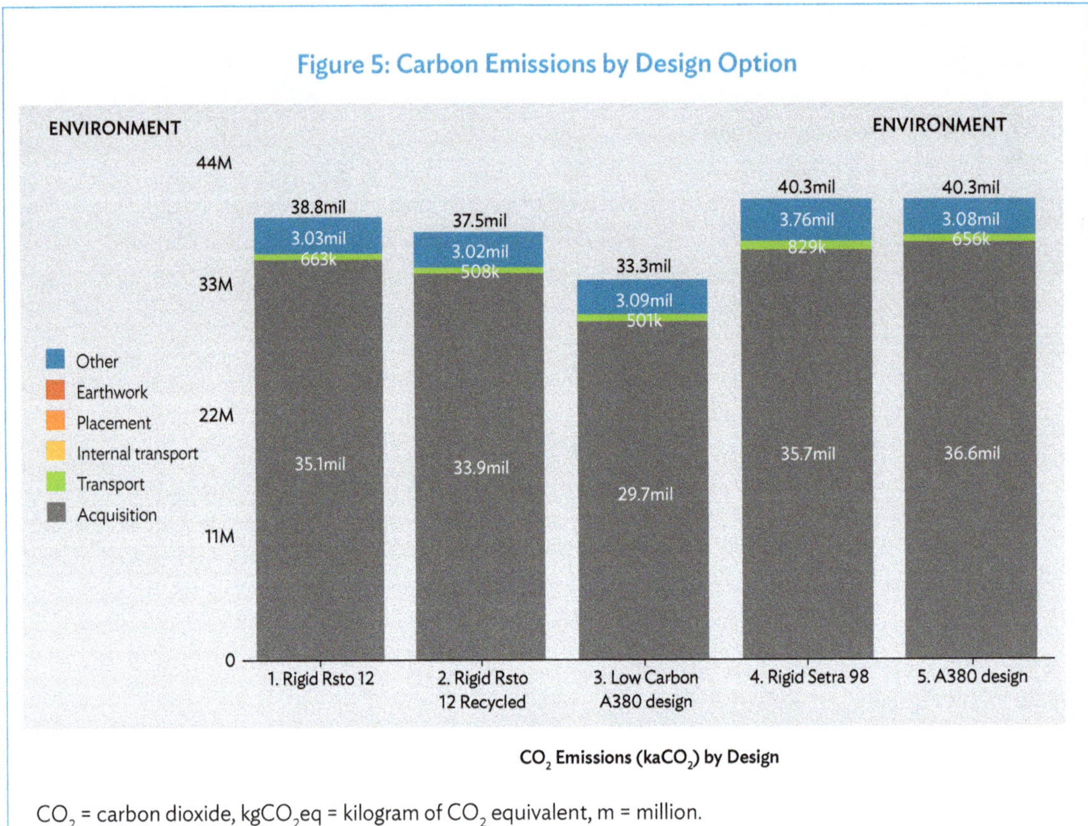

Figure 5: Carbon Emissions by Design Option

CO_2 = carbon dioxide, $kgCO_2eq$ = kilogram of CO_2 equivalent, m = million.

Source: ORIS. 2022. Digital Twin Road Solution Draft Report: Uzbekistan (A380 Section 1). Unpublished.

One of the objectives of the project was to reduce fatalities on this road section. Using the International Road Assessment Programme (iRAP) methodology, the assessment was able to develop a recommendation of measures, representing an investment of $7.1 million, to reach the 3-star minimum United Nations (UN) safety target and save over 1,800 injured persons and fatalities.

A. Assessment of Road Safety Using the International Road Assessment Programme Methodology

The iRAP Star Ratings are an objective measure of the level of safety, which is built-in to the road through more than 52 road attributes that influence risk for vehicle occupants, motorcyclists, bicyclists, and pedestrians. The ratings represent the relative risk of death and serious injury for an individual road user considering the following factors:

- **Likelihood.** Refers to road attribute risk factors that contribute to the risk of a crash being initiated;
- **Severity.** Refers to road attribute risk factors that account for the severity of a crash when it happens;
- **Operating speed.** Refers to factors that account for the degree to which risk changes with speed;
- **External flow influence factors.** Account for the degree to which a person's risk of being involved in a crash is a function of another person's use of the road; and
- **Median ability to traverse factors.** Account for the potential that an errant vehicle will cross a median (only applies to vehicle occupants and motorcyclists' runoff and head-on crashes).

All 52 attributes are recorded per 100-meter section of each road, which enables the calculation of the risks to vehicle occupants, motorcyclists, bicyclists, and pedestrians. The fatalities and serious injuries (FSIs) are then estimated by drawing on the road attribute data used for the iRAP Star Ratings, flow data for each road user, and network-level crash data to provide an estimation of FSIs along each segment of a road and support the prioritization of investments.

The ORIS platform integrates safety evaluation using the iRAP Star Rating and the FSI estimation methodology. The following assumptions were made for the A380 highway project:

- (i) Average speed: 120 km per hour (expected road limit),
- (ii) Road users: mainly vehicles (trucks) and motorcycles,
- (iii) User flow: 4,585 vehicle per day, and
- (iv) Fatality rate: 11 fatalities per year.[8]

The Star Ratings were below 3, which is lower than the UN minimum targets.

---

[8] Based on iRAP studies.

B. Countermeasure Plan

Building on this assessment and with the support of the ORIS team, ADB identified an improvement plan to reach the minimum iRAP Star Rating of 3, as required by both UN and ADB standards.

The countermeasure plan for the A380 highway section 673–698 km project consists of a mix of roadside barriers, shoulder rumble strips, and speed reductions at critical sections.

The implementation of these countermeasures increases the safety for all road users, thereby enabling the section to meet the targets of a minimum of 3 stars. The measures lead to a 56% reduction in the potential of FSIs, which would avoid 1,845 fatalities over the 30-year service life of the A380 highway section 673–698 km.

The additional investments for recommended measures are evaluated at $7.1 million, representing approximately 11% of the total project cost. Appendix 2 describes in more detail the road safety countermeasures.

Climate change is expected to impact the project over the next 40 years in terms of heat, freeze–thaw cycle, and water runoff floods. The analysis conducted with the ORIS platform identified a $19.4 million investment on adaptation and mitigation measures to limit damages and avoid early repair needs.

Resilience studies are integrating predictions of climate change and its impact on five parameters: heat, freeze–thaw cycle, runoff flooding, landslide, and silting. These parameters are being monitored between 2021 and 2060 by the Intergovernmental Panel on Climate Change (IPCC). The Shared Socioeconomic Pathways (SSP) are the emission scenarios driven by different socioeconomic assumptions. In the analysis, the selected scenario is the SSP5-8.5 (the worst-case scenario), which takes into account an earth warming rate of 8.5 watts per square meter, based on the world pollution trend proposed by the IPCC Sixth Assessment Report.[9] This path predicts a world that continues a fast-traditional development in emerging countries. The economic growth is based on a high consumption of fossil fuel and old technology with high production of CO_2eq.

The analysis conducted on the ORIS platform shows that the A380 highway 25 km section is at high risk of heat rise, heavy cycle of freeze–thaw, increase of water runoff, and silting. For each risk, a set of countermeasures was identified to mitigate those risks (Appendix 3). A template has also been created to assess the Paris Agreement alignment of the direct operations of the A380 highway upgrade (Appendix 4).

A. Heat

The heat risk is critically high for this road in some sections (Figure 6). The number of hot days is expected to increase from 148 days to 153 days per year, and the average temperature is projected to rise by 3.5°C in the next 40 years.[10]

With temperature of over 45°C at the surface of the asphalt concrete pavement, the resistance to rutting is decreased by threefold under traffic. Conversely, with surface temperature below 45°C, the structure resists three times more traffic than a pavement with temperature of over 45°C. Moreover, the rutting damage is increased with high surface temperature.[11] Overheating has a strong impact on the asphalt surface layer. As a black material, asphalt has a high capacity to absorb the heat, with a heat absorption rate of 0.80–0.95.

---

[9]   IPCC. Sixth Assessment Report. https://www.ipcc.ch/assessment-report/ar6/.

[10] This is according to predictions, based on the SSP5-8.5 IPCC model (IPCC. 2022. *Climate Change 2022: Impacts, Adaptation and Vulnerability*. Working Group II contribution to the IPCC Sixth Assessment Report. 28 February. https://www.ipcc.ch/report/ar6/wg2/).

[11] Z. A. Alkaissi. 2020. Effect of High Temperature and Traffic Loading on Rutting Performance of Flexible Pavement. *Journal of King Saud University - Engineering Sciences*. 32 (1). pp. 1–4. https://doi.org/10.1016/j.jksues.2018.04.005.

Figure 6: Heat Risk Modeling

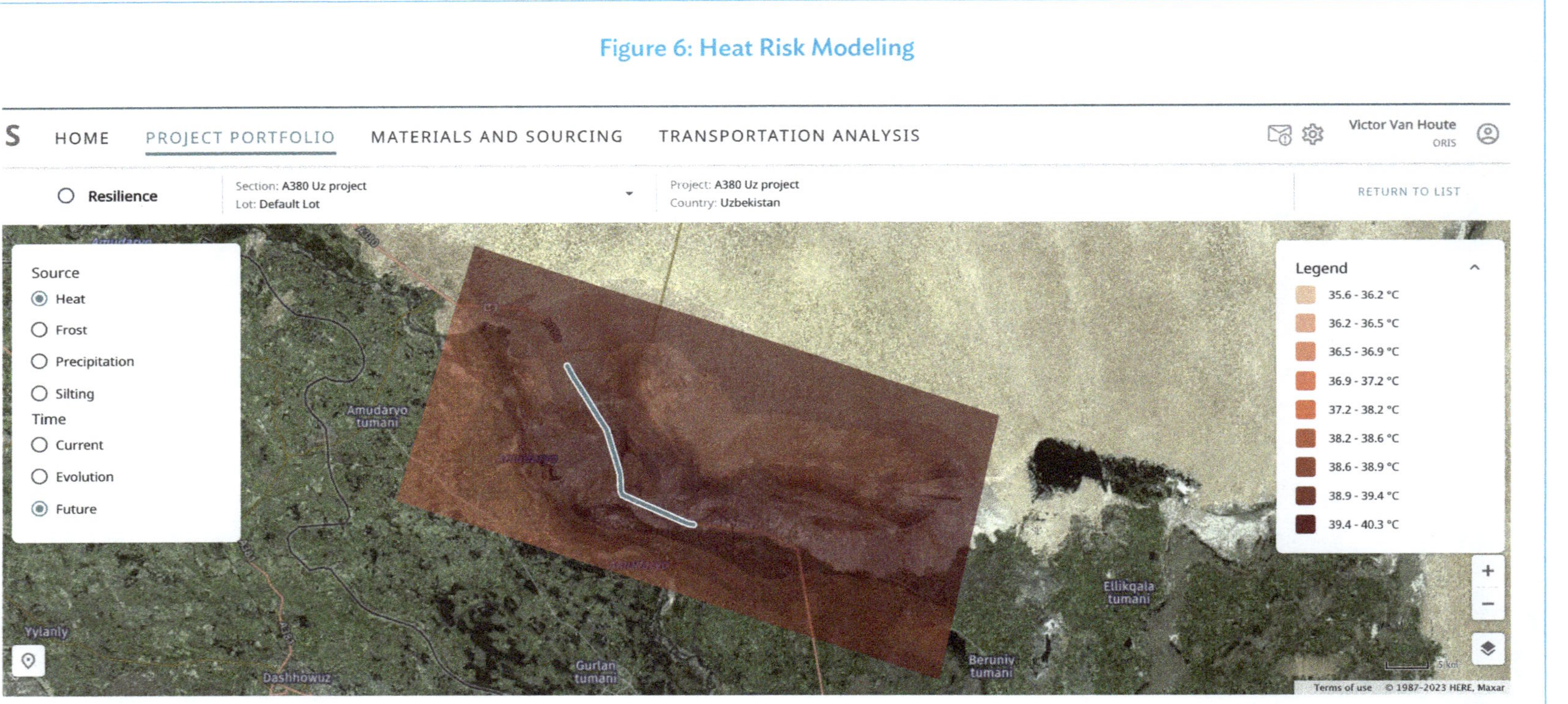

Source: ORIS. 2022. Digital Twin Road Solution Draft Report: Uzbekistan (A380 Section 1). Unpublished.

Therefore, the first adaptation measure is to confirm the initial choice of a concrete structure, the stiffness and behavior to overheating of which offer no sensitivity to the classical disease of overheating exposed asphalt networks. The A380 highway 25-km section built with a cement concrete surface layer will be structurally immune to this heat risk, bringing a longer service life to the infrastructure.

B. Freeze–Thaw Cycle

The average annual minimum temperature is expected to increase from the current 8.5°C to 11.5°C in 2060.[12] However, the number of freeze–thaw cycles is expected to vary from 58 for the historical period to 78 for 2060. Figure 7 illustrates the frost risk modeling.

The main impact of the freeze–thaw cycle on the cement rigid pavement is because of the weakening of the unbound granular materials and subgrades in road pavements during thaw periods in frost-affected regions, which leads to an early fatigue of the cement concrete surface layer. In addition, concrete slabs are also highly sensitive to frost heave, causing uneven movements of slabs that are very uncomfortable for drivers at the slab joint. Therefore, an adaptation measure consists of stabilizing the subgrade with a hydraulic binder on 30 centimeters, with a maximum rate of 6.0%.

C. Water Runoff

The water runoff hazard is strongly present in the center-south of the area (Figure 8), both for the current period and for the projection to 2060 (footnote 12). In addition, there is an increase in intensity of this hazard over the whole area. An adaptation measure is to resize the hydraulic structures.

D. Silting

The silting hazard is considered as high risk for the next 40 years (Figure 9), with the predicted reduction of rainfall identified as the root cause. Sand grains (with a diameter of 0.622 mm) are identified as the most impactful to prevent the silting risk. With the wind, these sand grains enter in motion by creeping and bouncing near the soil.

The planting of 147,000 saxaul trees is part of the project to mitigate the carbon impact. These trees could act as barriers as their height goes from 1 meter to 8 meters. Composed of thick trunks and many branches, these trees are well adapted to the local environment and will reduce the impact of sand silting.

---

[12] This is according to an ORIS Materials Intelligence analysis, based on SSP5-8.5 scenario.

Figure 7: Frost Risk Modeling

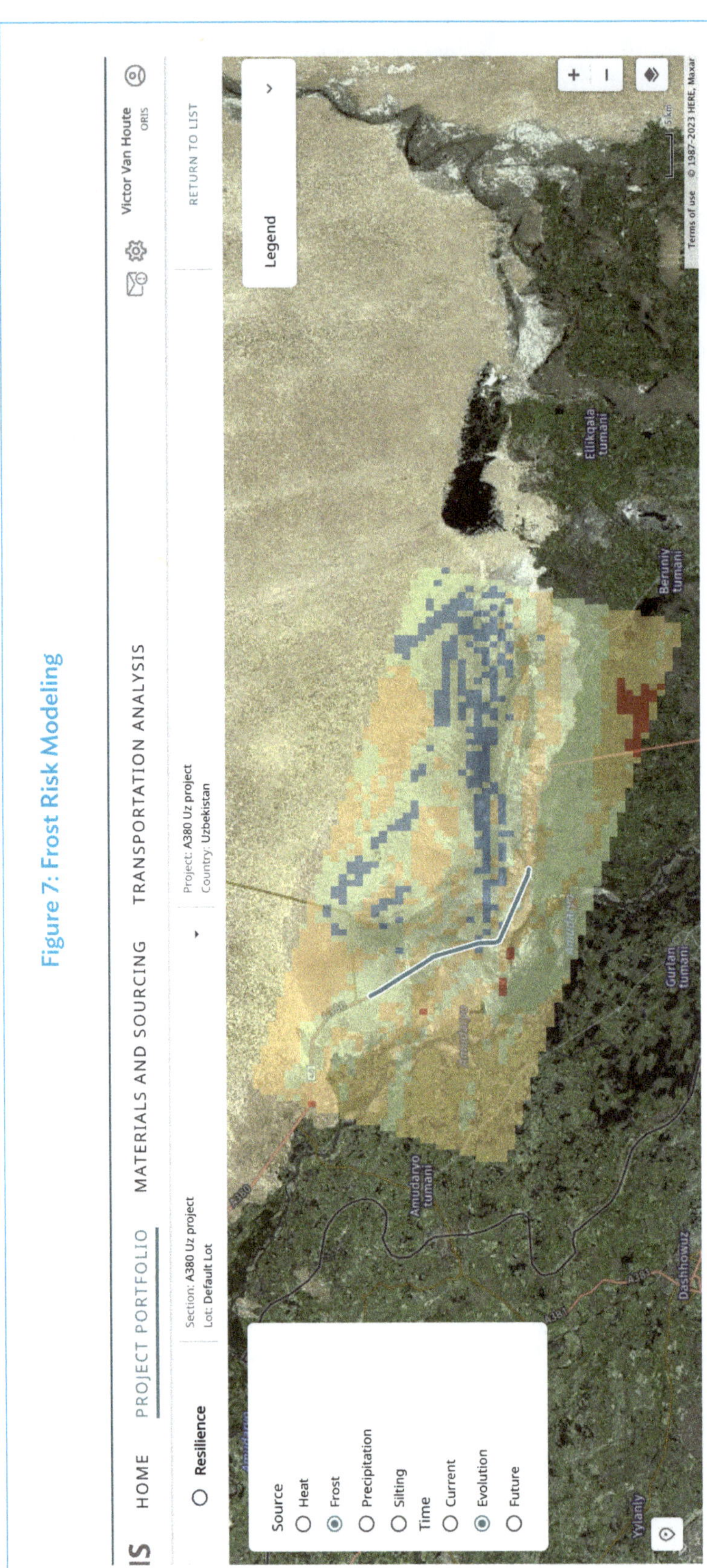

Source: ORIS. 2022. Digital Twin Road Solution Draft Report: Uzbekistan (A380 Section 1). Unpublished.

Figure 8: Runoff Risk Modeling

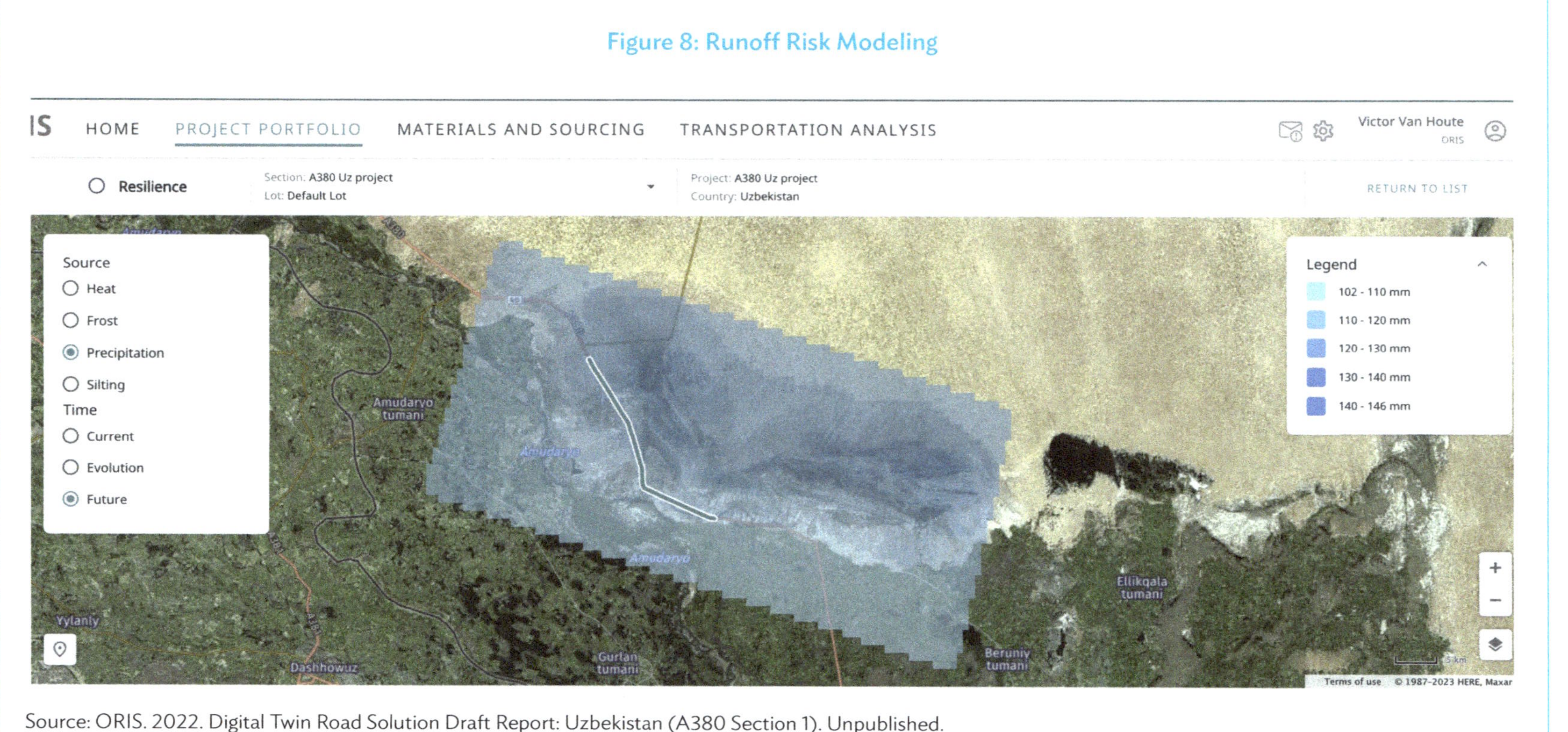

Source: ORIS. 2022. Digital Twin Road Solution Draft Report: Uzbekistan (A380 Section 1). Unpublished.

Figure 9: Silting Risk Modeling

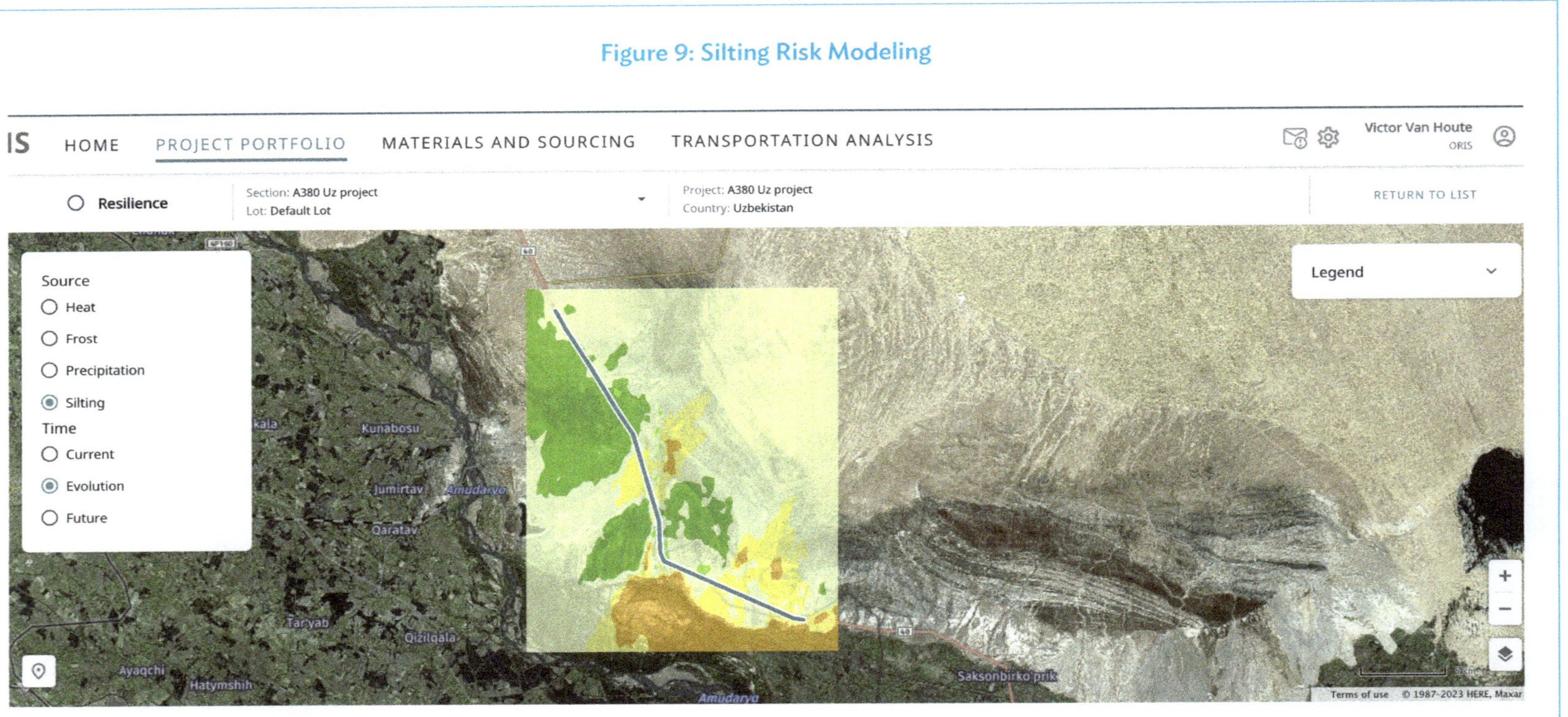

Source: ORIS. 2022. Digital Twin Road Solution Draft Report: Uzbekistan (A380 Section 1). Unpublished.

VII Conclusion and Key Learnings

The use of digital solutions offers great potential to invest in more sustainable, safe, and cost-efficient road networks.

This case study of the A380 highway section 673–698 km in Uzbekistan showcases that challenging initial designs can bring unexpected opportunities for improvement for the benefit of all. By creating a digital twin based on artificial intelligence, the ADB team was able to identify alternative designs that are more circular, with a lower carbon footprint and at a lower cost. It also allowed the project team to assess safety and resilience of the project and identify countermeasures.

Overall, the use of advanced digital tools allows identification of the optimum infrastructure design for a durable and long-term green investment, in line with the Sustainable Development Goals. Key learnings from this project are summarized in Box 2.

Box 2: Case Study on the Upgrade of Uzbekistan's A380 Highway—Key Learnings

The use of digital solutions, such as artificial intelligence and digital twins, to identify the optimum transportation infrastructure design resulted in key learnings, among which are the following:

- Using digital twin solutions to design transportation infrastructure offers great potential to build more sustainable infrastructure.
- Challenging initial designs within the existing norms with multiple iterations can help to identify the optimum design solution.
- Using advanced digital solutions allows the identification of long-term risks and countermeasures linked to climate change adaptation in order to build climate-resilient and durable transportation infrastructure.
- Thorough data collection on local sources of materials and their properties can help identify low-carbon alternatives for low-carbon designs.
- Analyzing the carbon footprint of projects through a life cycle assessment approach offers a comprehensive view and identification of key levers that will reduce the climate impact.
- Identifying circular solutions through these advanced digital approaches can help reduce the natural resources used for infrastructure transportation projects.
- Improving road safety and reaching the minimum requirement for the United Nations goals is possible by using a digital platform integrating international methodologies, such as the International Road Assessment Programme.

Digital due diligence and multiple-scenario analysis using the ORIS platform allowed the creation of quick iterations and the production of key outcomes in a shorter period of time than classical early stage studies. These two methods are complementary.

Source: ORIS. 2022. Digital Twin Road Solution Draft Report: Uzbekistan (A380 Section 1). Unpublished.

Appendix 1
Alternative Designs

Table A1: Comparison of the Design Alternatives with the Base Case

Pavement Design	Base Case Design	Alternative 1	Alternative 2	Alternative 2 with Recycled Layer	Base Case Design with Recycled Layer and Low-Carbon Products
Primary materials (kt)	909	18.8%	−1.5%	−9%	−29%
Water consumption (ML)	86.2	5.9%	−5.5%	−2.08%	0%
CO_2 (tCO_2eq)	39.8	−0.25%	−3.7%	−5.7%	−17%
Budget ($ million)	30.06	1.13%	−5.5%	−13.2%	−10.2%

kt = kiloton, ML = million liter, tCO_2eq = tons of carbon dioxide equivalent.
Source: ORIS. 2022. Digital Twin Road Solution Draft Report: Uzbekistan (A380 Section 1). Unpublished.

Road Safety Countermeasures

Table A2: Measures to Improve Safety

Countermeasure	Reduction of FSI	Justification	Cost ($ million)
Roadside barriers motorcyclist-friendly driver side	389 fatal and serious injuries over 20 years	Installation of 25 km of friendly motorcyclist barriers on the driver roadside brings safety for both vehicles and motorcyclist; moreover, those barriers create a safe area on the side of the road for pedestrian and avoid an additional accident once the pedestrian finds refuge on the other side of the barriers.	3.000
Roadside barriers motorcycle-friendly passenger side	320 fatal and serious injuries over 20 years	Add 20 km of barriers on the passenger side (already 5 km of road barriers has been registered in the drawings).	2.500
Shoulder rumble strips (both sides)	183 fatal and serious injuries over 20 years	Installation of rumble strips on the 25-km (both sides) project aims at reducing the frequency and severity of crashes due to driver performance errors. Installed along the edge of a travel lane, shoulder rumble strips produce noise and vibration that alert drivers when their vehicles are drifting off the roadway. This alert can also wake up drivers that fell asleep.	0.500
Treated shoulder on passenger side	74 fatal and serious injuries over 20 years	Shoulder stabilization on 1 m width and 30 cm depth.	0.090
Protected turn lane for intersection	33 fatal and serious injuries over 20 years	On four intersections (PK245, PK 186, PK 92, PK44), adding a protected turning lane to ensure a safe way out and into the A380 will increase the safety for road users.	1.000
Speed reduction	Motorcyclist rating raised from 3 to 4 stars	Intersections are the most critical area for motorcyclists in this project once the roadside barriers have been installed. The final improvement to increase the safety for motorcyclist and reach the UN target is to reduce the speed near the intersections from 100 to 90 km/h.	0.004
Total additionnal estimated costs ($ million)			**7.094**

Note: Numbers may not sum precisely because of rounding.

cm = centimeter, FSI = fatal and serious injuries, h = hour, km = kilometer, m = meter, UN = United Nations.

Source: ORIS. 2022. Digital Twin Road Solution Draft Report: Uzbekistan (A380 Section 1). Unpublished.

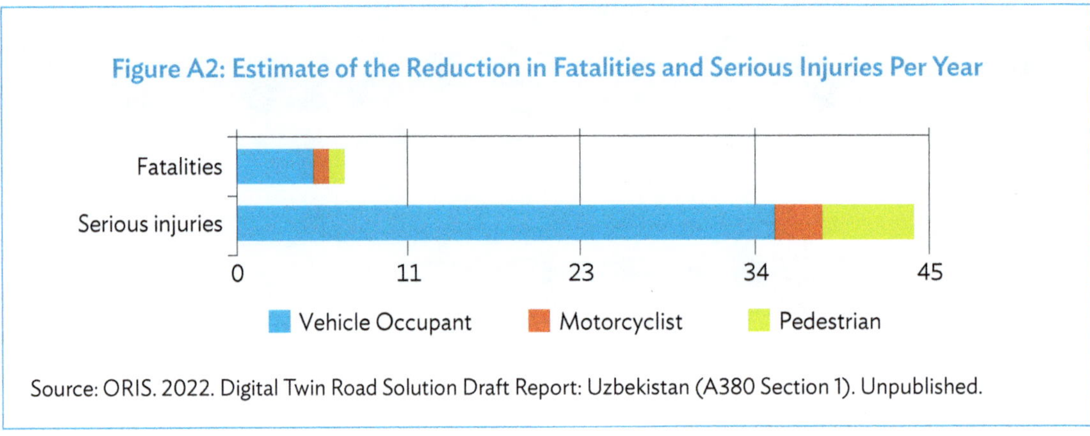

Source: ORIS. 2022. Digital Twin Road Solution Draft Report: Uzbekistan (A380 Section 1). Unpublished.

Appendix 3
Climate Risks

A380 Highway Climate Change Risk Assessment and Recommendations

Table A3: For a Resilient Project: Adaptation and Mitigation Recommendations

Risk	Evolution Prediction 2060 (SSP5-8.5)	Risk impact on the A380 Section 1	Adaptation/ Mitigation	Cost ($ million)	Countermeasure	Recommendation
Heat	The number of hot days will increase from 148 days to 153 days per year. Average air temperatures will increase by 3.5°C during the next 38 years.	High	Adaptation	1.32	Rigid pavement	Increase in temperature will have a high impact on a fully flexible pavement (for example, a rutting effect). Rigid pavement won't be affected by the risk and sustain the heat rise over the years. This adaptation is already planned in the project design.
Freeze–thaw cycle	Number of cycle freeze–thaw will increase by 33%. The intensity of frost will decrease.	High	Adaptation	1.53	Subgrade stabilization	Cycle of frost–thaw will impact the subgrade stability. Stabilization on 30 cm of this layer will bring a great strength to the pavement.
Water runoff	The total amount of waterfalls will decrease by 20% per month. However, the risk of runoff flooding will increase due to the flash rain increase and the soil capacity to stock water.	High	Adaptation	2.12	Culverts sizing control	Culvert sizing assessment taking predicted picks flow predicted in 2060. Resize the structures PK203+14, PK83+50. Create a structure at 42.0 70016, 60. 390414 Add extra drainages on critical points.

continued on next page

Table A3 *continued*

Risk	Evolution Prediction 2060 (SSP5-8.5)	Risk impact on the A380 Section 1	Adaptation/ Mitigation	Cost ($ million)	Countermeasure	Recommendation
Climate change	Taking appropriate action to minimize the CO_2 emissions	High	Adaptation	–3.20	Pavement design optimization	Optimize the pavement design with RStO 12 German alternative pavement. Recycle A380 materials as a granular road layer to reduce the CO_2 emissions of the project by 2.6 kt of $kgCO_2e$. Save five million liters of water.
Climate change	Taking appropriate action to minimize the CO_2 emissions	High	Mitigation	0.90	Tree planting	As trees grow, they absorb and store the CO_2 emissions that are driving global heating. Total pavement CO_2 emissions can be compensated with planting 58,100 trees (30 years of life).
Climate change	Industrial investment for low-carbon cement production	High	Mitigation	2.50	Financing industrial cement blender facility	The investment in industrial cement blender to produce and provide low-carbon cement to the A380 project will reduce the CO_2 emissions by 6.63 million $kgCO_2e$ and will keep producing low-carbon products for the other projects.
Landslide	The risk of landslide will increase on two-thirds of the project. However, the risk will remain low or moderate on the project	Low/ moderate	Adaptation	0.30	Monitor Landslide	No countermeasures to be implemented at construction stage. Recommendation: Monitor ground and slope conditions over time with drone survey during the 30 years of service life.

cm = centimeter, CO_2 = carbon dioxide, $kgCO_2e$ = kilogram of CO_2 equivalent, kt = kiloton.
Source: ORIS. 2022. Digital Twin Road Solution Draft Report: Uzbekistan (A380 Section 1). Unpublished.

Appendix 4
Paris Agreement Alignment

Template for the Paris Agreement Alignment Assessment of A380 Highway Upgrade Direct Operations

Disclaimer: This template was prepared in January 2022 based on the Guidance Note for the Paris Agreement Alignment Assessment V1. Therefore, results and conclusions displayed in this template might not be aligned with the latest guidance note available.

Table A4.1: Template for the Paris Agreement Alignment Assessment

1. General Information	
1.1 Name of person that prepared the case study	Hugo Pley-Leclercq (ORIS Project Manager)
1.2 Project name	A380 highway section 673–698 km (25-km)
1.3 Financing amount	$56 million
1.4 Economic sector(s)	Infrastructure
1.5 Economic subsector(s)	Transport (road)
1.6 Country, project location	Uzbekistan
1.7 Further information on the project	Upgrading and widening to four lanes of a 25-km section (673-698 km) of the Buzar–Qarshi–Bukhara–Nukus–Beyneu highway

2. Project Description

The existing 25-km section of the A380 highway will be upgraded. An existing section of the A380 highway (673-698 km) will be upgraded to a four-lane dual carriageway in line with traffic volumes and the standard applied in adjacent road sections. Cement concrete surfacing will be applied to increase resilience to higher temperatures, and increased drainage dimensions will be applied to take account for higher precipitation intensities due to climate change, ensuring a longer life span and reducing maintenance needs. The road will follow the existing alignment, and construction work will largely take place within the existing right-of-way. A road safety audit of the design will ensure appropriate measures are incorporated to avoid intersection-related crashes that constitute a significant proportion of road crashes in the country.

continued on next page

Table A4 *continued*

3. BB1 Assessment

Questions/ Assessment Steps	Answer and Justification	Clarifications (if assessing the project at project identification or concept paper stage)

Uniform Screening Criteria

#U1: Does the policy promote or support activities on the universally aligned list?	**NO** This project is not supporting activities on the universally aligned list. The A380 highway 25-km section project will increase the capacity of the A380 highway section with a rehabilitation for 2 X 1 carriageway to a separate 2 X 2 carriageway. The 1,204-km long A380 highway is part of the CAREC Corridor 2 and one of the key trade routes in the region, connecting the regions of Kashkadarya, Bukhara, Khorezm, and the Republic of Karakalpakstan and providing access to Kazakhstan and the Caspian Sea port of Aktau.	Preliminary comparison between the expected activities to be undertaken as part of the project and the universally aligned list. During project design, it will be necessary to monitor the initial comparison
#U2: Does the policy promote or support activities on the universally nonaligned list?	**NO** This transport project is not supporting the following activities: • Mining of thermal coal • Electric power generation from coal • Extraction of peat • Electricity from peat	

Specific Assessment Criteria

#SC1: Is the operation/economic activity inconsistent with the NDC of the country in which it takes place?	**NO** The project is not inconsistent with the NDC of the country. The NDC was updated in 2021, and states as its central goal the following: "The new goal of the Republic of Uzbekistan in terms of climate change mitigation, which seeks to be achieved by 2030, is hereby formulated as follows: reduce by 2030 specific greenhouse gas emissions per unit of GDP by 35% from the level of 2010." The document mentions the following transport-relevant implementation plans for reaching its NDC targets: "further introducing energy-saving technologies in industry, construction, agriculture, and other sectors of the economy," and "introducing alternative fuels in transportation." It is also mentioned that "The Ministry of Transport implements the gradual transition of public transport to natural gas and electric traction, and conducts measures to expand the production and use of vehicles with improved energy efficiency and environmental friendliness." The NDC does not stand in contradiction with building new road infrastructure, which is why the project is not considered inconsistent with SC1.	

CAREC = Central Asia Regional Economic Cooperation, GDP = gross domestic product, GHG = greenhouse gas, km = kilometer, LTS = long-term climate strategies, NDC = nationally determined contribution, UNFCCC = UN Framework Convention on Climate Change, WIM = weight in motion.

Source: Republic of Uzbekistan. 2021. *Updated Nationally Determined Contribution*. Tashkent. https://unfccc.int/sites/default/files/NDC/2022-06/Uzbekistan_Updated%20NDC_2021_EN.pdf.

#SC2: Is the operation/economic activity, over its lifetime, inconsistent with the country's LTS or other similar long-term national economy-wide, sector, or regional low-GHG strategies compatible with the mitigation goals of the Paris Agreement?

NO

Uzbekistan has not published an LTS. Based on a review of national climate- and transport-related strategies, it can be concluded that the project is not inconsistent with those. In fact, it directly supports the Strategy for the Development of the Transport System, as presented below.

The project also aims to establish a climate change committee or department under the Committee of Roads, with the objective to support the development and guide implementation of longer-term decarbonization pathways policy development, which could support the efforts of Uzbekistan in decarbonizing transport sector operations (as also presented below).

The strategy for the development of the transport system of the Republic of Uzbekistan until 2030 is based on seven main priority areas:
1. improving the efficiency of institutions and implementing a unified transport policy;
2. pursuing a balanced tariff policy, further reduction of the level of monopoly, and the formation of a healthy competitive environment in the market of transport services;
3. improving quality, accessible, and efficient transport services for the population and business;
4. deepening the integration of the transport system of the Republic Uzbekistan in the world transport space and implementation transit potential;
5. increasing the level of digitalization and introducing innovations in the transport system, creating an intelligent transport system;
6. improvement of the system for ensuring transport security and transportation safety; and
7. ensuring the environmental friendliness of transport, creating conditions for the development of "green" transport.

The project is aligned with priorities 3, 4, 5, 6, and 7 with
- Priority 3: Rehabilitation of the poor condition of section 673–698 km,
- Priority 4: Improve the A380 highway as part of the CAREC Corridor,
- Priority 5: Include WIM systems,
- Priority 6: Full analysis of the Safety and Recommendations to reach 3 stars rating and better, and
- Priority 7: Establishing a climate change committee to guide the development and implementation of pathways for decarbonizing the transport sector.

Uzbekistan's Third National Communication under the UNFCCC lists identified road sector-relevant measures for climate change mitigation in Table 4.1a Measures on Decrease in Energy Consumption of Automobile Transport. One of the measures includes the reconstruction of motor roads, which the project is fully aligned with.

continued on next page

Table A4.1a: Measures on Decrease in Energy Consumption of Automobile Transport

Items	Measures
Technical Measures	- further renewal of automobile, air, and railroad transport parks; - change over of automobile transport to run on liquefied or compressed natural gas; - organization of serial production in the country of automobile transport run on gas fuel; - construction of automobile gas refilling stations and workshops for reequipment of automobiles to run on gas fuel; - reconstruction and construction of motor roads; - further electrification of railroads; - public transport traffic optimization in large cities of the republic; - introduction of hybrid electrical automobile transport; - quality improvement of engine fuel and development of new types of engine fuel; - carrying out "clean air" campaigns.
Institutional Measures	- introduction of fuel consumption standards; - introduction of tires marking; - establishment of CO_2 emission standards; - "modal shift" or priority development of urban public transport, including access - limitation to cities center, establishment of paid parking, development of bicycle infrastructure; - establishment of system for metering energy resources consumption in "Transport" Sector.

The Strategy on the Transition of the Republic of Uzbekistan to a "Green" Economy for the Period 2019–2030 mentions expanding the production and use of motor fuels and vehicles with improved energy efficiency and environmental friendliness, as well as the development of electric transport. According to the NDC, the Environmental Protection Concept-2030 (UP-5863 dated 30.10.2019) mentions transition of 80% (about 6,500) of public transport units to gas fuel and electric traction.

Sources:
Strategy for the Development of the Transport System of the Republic of Uzbekistan until 2030. https://unece.org/sites/default/files/2021-07/A.Muminov-Ministry-of-transport-rus.pdf.

Third National Communication of the Republic of Uzbekistan under the UN Framework Convention on Climate Change. https://unfccc.int/sites/default/files/resource/TNC%20of%20Uzbekistan%20under%20UNFCCC_english_n.pdf.

#SC3: Is the operation/economic activity inconsistent with global sector-specific decarbonization pathways in line with the Paris Agreement mitigation goals, considering countries' common but differentiated responsibilities and respective capabilities?

NO

The project is not considered inconsistent with global decarbonization pathways. Most global pathways (including those cited in the IPCC's latest Sixth Assessment Report Working Group III contribution and IEA's Net Zero by 2050) require a transition to e-mobility, induced vehicle efficiency, as well as modal shifts (e.g., from public to private transport) within the upcoming decades, up to 2050. Developed countries are expected to be the first to transition, whereas the decarbonization transition in developing countries will require longer time periods, depending on local capacity, capability, and availability of resources. Road infrastructure will be required for supporting also decarbonized modes of transport, which provides one justification for this project not being inconsistent with global pathways.

continued on next page

In addition, the project will aim to support a transition to decarbonized transport operations in Uzbekistan through the following aspects:

- It will set up a climate change committee/department within the Committee of Roads (under the Ministry of Transport). The aim of this committee/department is to support the development and implementation of a decarbonization pathway, or equivalent, for the road sector. In its NDC, Uzbekistan mentioned that it will "strive to formulate and communicate a long-term low GHG development strategy based on its own national circumstances." This committee will support this task from a road subsector perspective. ADB plans to develop a TA for Uzbekistan for supporting this work. The country has shifted its LDV fleet from gasoline to natural gas from 15% in 2015 to 60% in 2022, with a mitigating impact on GHG emissions (natural gas reduces GHG emissions with approximately 20% compared to gasoline). Still, further policy work and implementation are needed for ensuring decarbonization of the sector.
- The project includes repaving access roads to railway connections, which directly support multimodal integration between road transport and the lower-carbon mode rail. It is expected to spur multimodal logistics, including cement transportation using rail. The road also provides access to the closest port, and can thus be seen as a (minor) supporting link to multimodal maritime transport/logistics.

Further, in terms of direct project emissions relating to construction works, the project is designed to reduce its carbon impact with the following mitigation measures:

- The pavement structure benefits from a recycled layer (165,000 tons), contributing to carbon savings and low-carbon material industrial production of 18% (6.79 million $kgCO_2e$);
- Planting of 147,000 saxaul trees;
- The new ADB procurement guideline will be used to encourage contractors to use e-trucks and low-carbon emissions equipment to reduce the carbon emissions link to the material transport; and
- Consulting engineers proposed to include in the project an investment for an industrial cement blender. This blender could provide lower-carbon cement (from 35% to 75%) for all cement-related constructions.

Finally, based on HDM-4 modeling, there is potential for the road to decrease vehicle-specific transport operations emissions by allowing for more stable vehicle speed throughout the project life cycle of 30 years.

Sources:
- 2011–2015 program, 15% of the vehicles: UZ Daily. 2011. Number of Cars, Consuming Compressed Gas, to Reach 29% in Uzbekistan by 2016. 17 January. https://www.uzdaily.uz/en/post/12839.
- 2022, 60%: Eurasianet. 2022. Uzbekistan Ends Gas Exports to China, Abandons Price Increases at Home. 10 January. https://eurasianet.org/uzbekistan-ends-gas-exports-to-china-abandons-price-increases-at-home.

#SC4: Does the operation/economic activity prevent opportunities to transition to Paris-aligned activities, or primarily support or directly depend on nonaligned activities in a specific country/sector context?

NO

The project does not prevent opportunities to Paris-aligned activities and does not support nonaligned activities. While new infrastructure is constructed that might induce traffic, new road infrastructure will be required for allowing the transition to decarbonized road transport. At the same time, the project takes a concrete step toward supporting the development of a longer-term decarbonization pathway for the road sector through the establishment of the climate change committee/department within the Committee of Roads. This way, the objective is to support a longer-term transition toward decarbonized road transport, with an impact on future emissions of road operations. Further, the new infrastructure provides better multimodal transportation with connections with train stations. Thus, it does not prevent opportunities for the development of low-carbon transportation options; instead, it supports multimodal rail logistics.

#SC5: Is the operation/economic activity economically unviable, when taking into account the risks of stranded assets and transition risks in the national/sector context?

NO

Total estimated carbon (construction + vehicles running + vehicles manufacture): 3,371,000 tCO_2e

Social cost of carbon = $13,484,000 ($40/tCO_2e)

The economic and financial analyses have been done. Results are presented below:
- The results of the economic analysis are summarized in Table A4.1b, expressed in terms of the key economic indicators: benefit–cost ratio, EIRR, and net present value at a 9% discount rate. The overall project has an EIRR of 13.1% and is therefore economically viable, and both A380 highway sections have an EIRR well above 9.0%. The rural road section gives an EIRR below 9% but well above the 6% threshold considered acceptable for projects serving basic needs of rural communities and also providing access to the new mining and metallurgical center.
- "The project is nonrevenue generating, so the objective of the analysis is to ensure that it is financially sustainable. With a new concrete pavement, the project is expected to reduce the periodic and routine maintenance requirements considerably. The HDM-4 analysis used for the economic evaluation estimated an annual without-project cost of about $0.49 million per annum, falling to about $0.45 million in the with-project case."

continued on next page

Table A4 *continued*

Table A4.1b: Results of Economic Analysis

Project Road	EIRR (%)	NPV ($m)	Benefit-Cost Ration
New 25-km A380 (Km 673 to Km 698)	17.4	77.6	3.0
New 4R180 Section (4 Km Rural road section)	8.3	-0.2	0.9
Original project with cost revised (A380 [Km 964 to 1,204])	12.4	162.5	1.6
Overall Project	**13.1**	**239.9**	**1.8**

EIRR = economic internal rate of return, km = kilometer, NPV = net present value.
Note: NPV uses a 9% discount rate.

Result	**The project A380 highway** section 673–698 km **is aligned as per the BB1.**

4. BB2 Assessment

Questions/ Assessment Steps	**Answer and Justification Clarifications** (if assessing the project at project identification or concept paper stage)

CRITERION 1: Establishment of Climate Risk and Vulnerability Context

| **Step 1:** Identifying and assessing physical climate risk—Is the operation (including assets, stakeholders, and the system within which it takes place) at risk? | **YES**

1. Is the project exposed and vulnerable to the impacts of climate change within its spatial and temporal boundaries? **YES**

Climate risk assessment studies have been made by ORIS Materials Intelligence. The risks categories identified are the heat evolution, freeze–thaw cycle, water runoff, silting, and landslide.

The table below extracted from the consultant report grades those risks from *low* to *high*.

2. Is the overall climate risk within the operation boundary? **YES**

The study shows that the impact of heat, freeze–thaw cycle, water runoff, and silting are *high* for the next 40 years because of climate change:
• The number of hot days will increase from 148 to 153 per year. Average air temperatures will increase by 3.5°C during the next 38 years. Increase in temperature will have a high impact on a fully flexible pavement (e.g., a rutting effect).
• The number of freeze–thaw cycles will increase by 33%. The intensity of frost will decrease. Freeze–thaw cycle will impact the subgrade stability. | See ORIS–ADB Final Report A380 highway section 673–698 km:

5.5 Resilience to Climate Change |

continued on next page

Table A4 *continued*

- The total amount of waterfalls will decrease by 20% per month. However, the risk of runoff flooding will increase because of the increase in flash rain and the poor soil capacity to stock water.
- Silting effect will increase with the decrease of waterfalls. The silting effect increases the risk of accidents for road users.

CRITERION 2: Definition of the Climate Adaptation and Resilience Measures

Step 2: Addressing physical climate risks and building climate resilience—Have climate adaptation and resilience measures been identified to reduce material physical climate risks and contribute to building climate resilience?

YES

Adaptations and mitigation investment plans at construction stage have been proposed (see Appendix 3). The investment cost at construction stage for adaptation and mitigation measures is estimated at $19,410,000 (around 34.7% of the total project cost).

These adaptation and mitigation measures will increase the resilience of the road against climate change in the next 40 years.

1. Which adaptation measures have been incorporated in the project design to address the identified climate risks?

   These adaptation measures include
   - Use of rigid pavement structure (instead of flexible pavement);
   - Reinforcement of the subgrade, thanks to the stabilization with hydraulic binder on 30 cm depth;
   - Resizing of the drainage system, increasing the slopes protection, and increasing the size of some culverts; and
   - Planting of 147,000 saxaul trees.

2. Does the project have an opportunity to enhance climate resilience? **YES**
   - Use of rigid pavement will improve the climate resilience of the project.

3. Could the proposed climate adaptation and resilient measures contribute to maladaptation? **NO**

4. Have the climate adaptation and resilience measures selected for the project been documented? **YES**, measures are compiled in the consulting report.

See ORIS–ADB Final Report A380 highway section 673–698 km:

5.5 Resilience to Climate change

6.3. Verification of the main hydraulic structures

6.5. Countermeasures, adaptation, and mitigation

6.6. Mitigation measures to reduce the impact of climate change

6.7. Conclusion: Climate change countermeasures

Table A4 *continued*

CRITERION 3: Assessment of Inconsistency with a National/Broad Context for Climate Resilience

Step 3: Assessing the broader climate resilience context—Is the operation not inconsistent with relevant national policies/strategies, private sector or community-driven priorities for climate adaptation and resilience?

YES

1. Which policies and strategies for climate adaptation and resilience exist and are relevant for the project? **NONE**

   Main policies for the country are based on mitigation of climate change with refurbishment/reconstruction of the road.

   There is no NDC mentioning adaptation for road networks in Table A4.1c Measures on Decrease in Energy Consumption of Automobile Transport found in the Uzbek NDC n°3. The reconstruction of motorways is promoted as an NDC measure.

Table A4.1c: Measures on Decrease in Energy Consumption for the Road Sector

Items	Measures
Technical Measures	- further renewal of automobile, air, and railroad transport parks; - change over of automobile transport to run on liquefied or compressed natural gas; - organization of serial production in the country of automobile transport run on gas fuel; - construction of automobile gas refilling stations and workshops for reequipment of automobiles to run on gas fuel; - reconstruction and construction of motor roads; - further electrification of railroads; - public transport traffic optimization in large cities of the republic; - introduction of hybrid electrical automobile transport; - quality improvement of engine fuel and development of new types of engine fuel; - carrying out "clean air" campaigns.
Institutional Measures	- introduction of fuel consumption standards; - introduction of tires marking; - establishment of CO_2 emission standards; - "modal shift" or priority development of urban public transport, including access - limitation to cities center, establishment of paid parking, development of bicycle infrastructure; - establishment of system for metering energy resources consumption in "Transport" Sector.

2. Is the project, overall, not inconsistent with priorities identified in these policies/strategies? **NO**

   Renovation of the road will reduce the use of fuel per vehicle.

Result

The A380 highway section 673–698 km **is aligned as per the BB2.**

Source: Centre of Hydrometeorological Service at Cabinet of Ministers of the Republic of Uzbekistan (Uzhydromet). 2016. *The Third National Communication of the Republic of Uzbekistan under United Nations Framework Convention on Climate Change.* Tashkent. https://unfccc.int/sites/default/files/resource/TNC%20of%20Uzbekistan%20under%20UNFCCC_english_n.pdf.

5. Paris Agreement Alignment Assessment Result

The project A380 highway section 673–698 km **is aligned with the Paris Agreement**. The following assessments have been made:

BB1 Assessment: Aligned
- U1: NO
- U2: NO
  - » SC1: NO
  - » SC2: NO
  - » SC3: NO
  - » SC4: NO
  - » SC5: NO

BB2 Assessment: Aligned
- Criterion 1: YES
- Criterion 2: YES
- Criterion 3: YES

ADB = Asian Development Bank, CAREC = Central Asia Regional Economic Cooperation, cm = centimeter, EIRR = economic internal rate of return, GDP = gross domestic product, GHG = greenhouse gas, HDM = Highway Development and Management, IEA = International Energy Agency, IPCC = Intergovernmental Panel on Climate Change, $kgCO_2e$ = kilogram of carbon dioxide equivalent, km = kilometer, LDV = light duty vehicle, LTS = long-term strategy, NDC = nationally determined contribution, tCO_2e = ton of carbon dioxide equivalent, UNFCCC = United Nations Framework Convention on Climate Change, WIM = weigh-in-motion.

Source: ORIS. 2022. Digital Twin Road Solution Draft Report: Uzbekistan (A380 Section 1). Unpublished.

Akbarian, M. 2012. Model Based Pavement-Vehicle Interaction Simulation for Life Cycle Assessment of Pavements. Doctoral thesis. Cambridge: Massachusetts Institute of Technology.

CEN. 2016. Sustainability of Construction Works - Environmental Product Declarations - Product Category Rules for Concrete and Concrete Elements. EN 16757: 2017 (Main).

Hajj, E. Y. et al. 2017. *Enhanced Prediction of Vehicle Fuel Economy and Other Vehicle Operating Costs, Phase I: Modeling the Relationship Between Vehicle Speed and Fuel Consumption.* Technical report prepared for the US Department of Transportation Federal Highway Administration. Reno: University of Nevada.

Hajj, E. Y. et al. 2018. *Enhanced Prediction of Vehicle Fuel Economy and Other Vehicle Operating Costs. Task 9: Incremental Fuel Consumption Due to Pavement Roughness. Appendix B: Fuel Economy and Fuel Consumption Results.* Technical report prepared for the US Department of Transportation Federal Highway Administration. Reno: University of Nevada.

Hajj, E. Y. et al. 2018. *Enhanced Prediction of Vehicle Fuel Economy and Other Vehicle Operating Costs. Task 9a–b: Incremental Fuel Consumption Due to Pavement Roughness.* Technical report prepared for the US Department of Transportation Federal Highway Administration. Reno: University of Nevada.

Harvey, J. et al. 2016. Simulation of Cumulative Annual Impact of Pavement Structural Response on Vehicle Fuel Economy for California Test Sections. *UCPRC Research Report. No. UCPRC-RR-2015-05.* Berkeley: University of California Pavement Research Center (UCPRC). https://escholarship.org/content/qt3p8312vs/qt3p8312vs_noSplash_83d42ade4d3fcc6214355310d59e6579.pdf.

Xu, X. G. 2017. *Evaluation of the Albedo-Induced Radiative Forcing and CO2 Equivalence Savings: A Case Study on Reflective Pavements in Four Selected U.S. Urban Areas.* Cambridge: Massachusetts Institute of Technology.

Zaabar, I. C. 2012. Estimating the Effects of Pavement Condition on Vehicle Operating Costs. *NCHRP Report 720.* Washington, DC: Transportation Research Board of the National Academies.

www.ingramcontent.com/pod-product-compliance
Lightning Source LLC
LaVergne TN
LVHW071455180726
843512LV00018B/1391